Riding Mail Trail

Contents

Chapter 1

Staying in Touch

How many ways do you talk with friends and family? Even if friends live a long way away, it is easy to still talk with them.

E-mail is a way to send messages. Some people send instant messages on their computers or text messages on their cell phones. Even letters travel quickly.

It was not always this easy to share information.

Did You Know?

There are twice as many e-mails sent now as in 2003!

The U.S. Postal Service delivered about 213 billion pieces of mail to more than 250 million homes and 8 million businesses in 2007.

During the 1800s, people started moving west. The Gold Rush brought large numbers of people hoping to become rich. By 1860, more than 1,000,000 people had moved west to California, Kansas, Nebraska, Nevada, Oregon, and Texas.

U.S. Population by Region – 1860

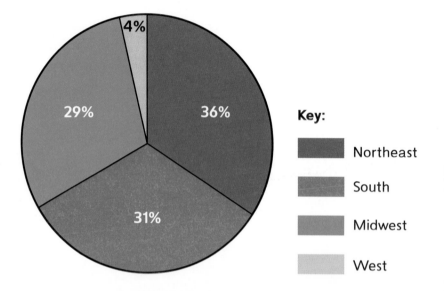

Key:

- Northeast
- South
- Midwest
- West

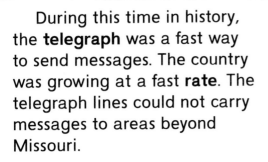

During this time in history, the **telegraph** was a fast way to send messages. The country was growing at a fast **rate**. The telegraph lines could not carry messages to areas beyond Missouri.

The telegraph used Morse code, a series of dots and dashes, to send messages. Operators had to learn the code. Using a maximum of 3 dots or dashes, how many letters could be made?

One way mail could reach the West Coast was by boat. This trip took a long time. A letter from the east had to travel by ship to Panama. It crossed Panama by **stagecoach** or train.

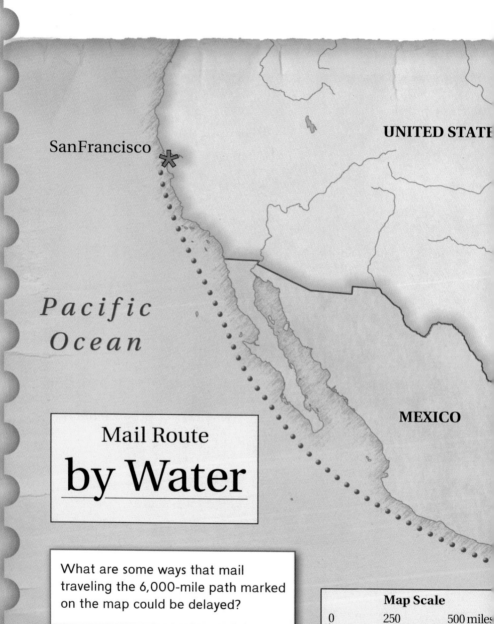

SanFrancisco

UNITED STATE

Pacific Ocean

MEXICO

Mail Route
by Water

What are some ways that mail traveling the 6,000-mile path marked on the map could be delayed?

Map Scale

| 0 | 250 | 500 miles |

The letter was put on another ship headed for California. The journey could take several months. The ship would dock in only one place on the West Coast. That was usually at a port in San Francisco. How might this have been a problem?

New York

N
W   E
S

Atlantic Ocean

lf of Mexico

Caribbean Sea

Isthmus of Panama

PANAMA

There were also 2 overland mail routes. The mail traveled west by stagecoach. This trip was supposed to take about 24 days, but sometimes it took months. What are some possible reasons for the delay?

OVERLAND MAIL ROUTE
TO CALIFORNIA.

Through in Six Days to Sacramento!

CONNECTING WITH THE DAILY STAGES
To all the Interior Mining Towns in Northern California and Southern Oregon.
Ticketed through from Portland, by the

OREGON LINE OF STAGE COACHES!
And the Rail Road from Oroville to Sacramento,
Passing through Oregon City, Salem, Albany, Corvallis, Eugene City, Oakland,
Winchester, Roseburg, Canyonville, Jacksonville, and in California—
Yreka, Trinity Centre, Shasta, Red Bluff, Tehama, Chico,
Oroville, Marysville to Sacramento.

The Overland Mail Company charged 10¢ for each letter. It also carried passengers at a charge of 10¢ per mile.

HOLLADAY OVERLAND MAIL & EXPRESS CO

It took a long time for news to travel across the country. People needed a faster way to send news.

The founders of the Pony Express hoped that their way would be the fastest. The government was going to give $1 million to the company with the best idea for delivering the mail.

The plan worked. In only 9 days, mail made the almost 2,000-mile trip from St. Joseph, MO, to Sacramento, CA. The Pony Express was in business.

Did You Know?

The $1 million mail contract would be worth over $20.5 million in today's money.

Chapter II

The Trail West

On April 3, 1860, a young rider named Johnny Fry got on his horse in St. Joseph, MO. Early the next morning, Billy Hamilton left Sacramento, CA, riding east. The two riders would meet halfway. They would exchange bags of mail. Then they would turn around and go back the way they came.

TALK ABOUT IT

Compare the Pony Express to a relay race.

OVERLAND MAIL ROUTE
TO CALIFORNIA.
Through in Six Days to Sacramento!

Johnny Fry rode from St. Joseph, MO, to Seneca, KS, at about 12 miles per hour.

About 160 stations were set up all along the route. They were located 10 to 12 miles apart.

The horses could run about 10 miles per hour. Horses traveled no more than 15 miles in a route. Riders used a new horse on each part of the trip.

At 10 miles an hour, about how many hours did the 15-mile routes take?

Route of the

Pony Express

1860–1861

Sacramento

Carson City

Salt Lake City

NEVADA

UTAH

WYOM

CALIFORNIA

Pacific

Ocean

N

W   E

S

— Pony Express route

When a Pony Express rider reached a station, he changed horses. He put his bag of mail, called a *mochila* (moh CHEE luh), on the new horse. Then he went on down on the trail.

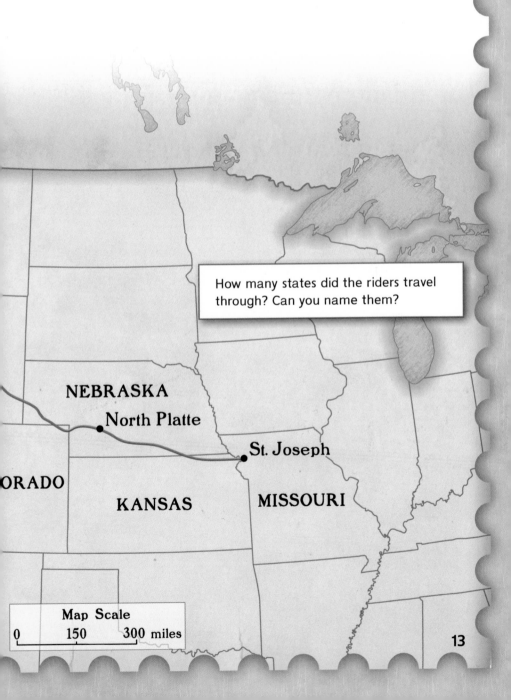

How many states did the riders travel through? Can you name them?

NEBRASKA

North Platte

St. Joseph

ORADO

KANSAS

MISSOURI

Map Scale

0   150   300 miles

Chapter III

Surviving the Ride

Weight was important. The horses could keep up their speed as long as they carried 165 pounds or less. *Mochilas* could hold about 20 pounds of mail. The horse also carried another 20 pounds of equipment. This meant that riders had to weigh less than 125 pounds.

Most riders were thin young men. The youngest rider, Bronco Charlie Miller, was only 11 years old. Most of the other riders were between 18–25 years old.

< 125 pounds

20 pounds of mail +
20 pounds of other equipment

Total Weight ≤ 165 pounds

Each rider rode a route that was 75–100 miles long. Every third station was set up as a home station. Which state had the most home stations?

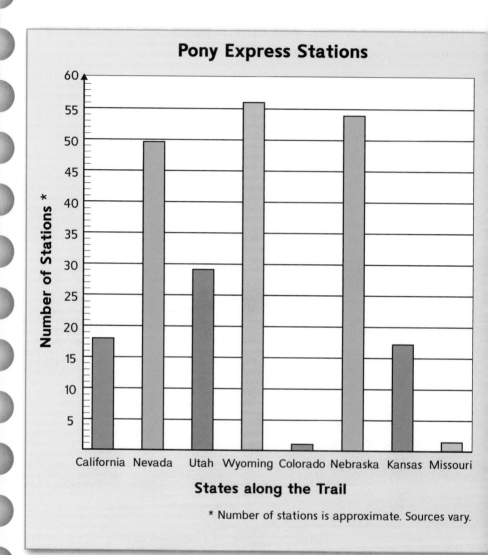

Pony Express Stations

Number of Stations *

States along the Trail

* Number of stations is approximate. Sources vary.

Riders traveled through large parts of some states and barely into other states along the trail.

Home stations had extra horses, food, and other items they might need. A rider would rest after passing the mail to the next rider.

A Pony Express home station.

The brave men who rode the Pony Express had a dangerous job. They rode through mountains, valleys, and deserts. They could run into thieves or wild animals. They also could get lost in a blizzard or run out of water in the desert.

In spite of all the risks, only one rider died on the trail.

TALK ABOUT IT

The Pony Express riders earned about $100 a month. That would be equal to about $2,000 a month today. Would you have been a Pony Express rider? Explain.

Important events before and during the time of the Pony Express

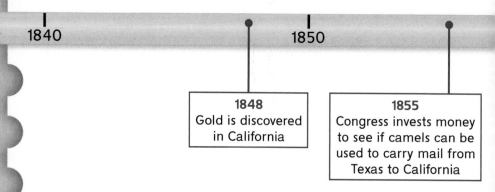

1840   1850

1848
Gold is discovered in California

1855
Congress invests money to see if camels can be used to carry mail from Texas to California

After only 1½ years, the Pony Express was no longer needed. By 1861, telegraph lines stretched from coast to coast. The telegraph was the fastest way to send information.

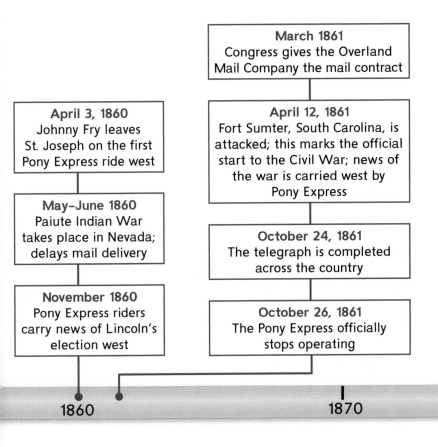

March 1861
Congress gives the Overland Mail Company the mail contract

April 3, 1860
Johnny Fry leaves St. Joseph on the first Pony Express ride west

April 12, 1861
Fort Sumter, South Carolina, is attacked; this marks the official start to the Civil War; news of the war is carried west by Pony Express

May–June 1860
Paiute Indian War takes place in Nevada; delays mail delivery

October 24, 1861
The telegraph is completed across the country

November 1860
Pony Express riders carry news of Lincoln's election west

October 26, 1861
The Pony Express officially stops operating

1860   1870

Chapter IV

Just a Moment in History

During its 18 months, the Pony Express delivered about 35,000 pieces of mail. Only one bag of mail was ever lost.

Still, the Pony Express was not seen as a success. When the Pony Express ended, its founders had no money. They had **invested** $700,000, and were $200,000 in **debt**.

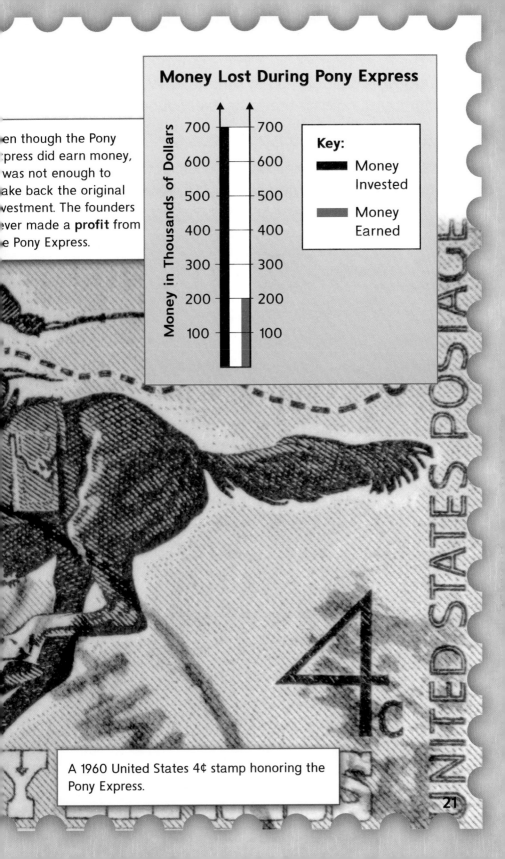

en though the Pony
press did earn money,
was not enough to
ake back the original
vestment. The founders
ever made a **profit** from
e Pony Express.

Money Lost During Pony Express

Money in Thousands of Dollars

700
600
500
400
300
200
100

700
600
500
400
300
200
100

Key:

████ Money Invested

▓▓▓▓ Money Earned

A 1960 United States 4¢ stamp honoring the Pony Express.

The Pony Express is an important piece of our country's history. It is part of the old American West. Members of a historic group rerun the route every June.

Just like the original riders, the riders travel west on horseback carrying mail in a *mochila*. Today's riders carry letters that can be purchased by collectors. The envelopes show that they traveled by Pony Express.

About 10 days later, they deliver the mail to California. More than 500 people take part in the event. This is their way of honoring this part of American history.

Glossary

debt
> Owing money to someone. *(page 20)*

invested
> Past tense of *invest*; to put money into something with the hope that it will result in a profit. *(page 20)*

profit
> Amount of money made by a business that exceeds the amount put in at the start. *(page 21)*

rate
> A measurement of one thing in terms of something else. *(page 5)*

stagecoach
> A horse-drawn mail carriage. *(page 6)*

telegraph
> A device used to communicate by coded signal. *(page 5)*